RAISED-BED GARDENING FOR BEGINNERS

A step by step guide to grow outdoor plants and vegetables

Lily Mercer

Copyright

Table of Contents

Introduction

As a child, I spent countless hours exploring my grandmother's garden, a wonderland of colorful flowers, fragrant herbs, and bountiful vegetables. I was fascinated by the magic of transformation, watching tiny seeds sprout into thriving plants that yielded delicious fruits and nourishing greens.

Years later, as an adult yearning for a connection to nature and a source of wholesome food, I rediscovered the enchanting world of gardening. Inspired by the concept of raised beds, I envisioned creating my own personal oasis, a place where I could nurture life from seed to harvest.

With a mix of excitement and trepidation, I embarked on my raised-bed gardening journey. I eagerly delved into research, seeking knowledge from experienced gardeners and experts. I experimented with different techniques, learning from my successes and failures.

Through hands-on experience, I discovered the joys and challenges of raised-bed gardening. I learned the importance of soil composition, the art of selecting suitable plants, and the delicate balance of nurturing and protection.

With each passing season, my raised beds transformed into a thriving ecosystem, a testament to the power of nature and the rewards of patient care. I reveled in the satisfaction of harvesting fresh, sun-ripened fruits and vegetables, savoring their flavors and sharing them with loved ones.

Raised-bed gardening became more than just a hobby; it evolved into a passion, a way of life that instilled in me a deep appreciation for the natural world. I discovered that gardening is not merely about cultivating plants; it's about nurturing life, fostering harmony with nature, and finding joy in the simple act of growing something from seed to harvest.

In this comprehensive guide, I invite you to join me on a journey of discovery, where we'll explore the wonders of raised-bed gardening together. We'll delve into the intricacies of soil preparation, plant selection, and ongoing care, transforming our backyards into flourishing havens of homegrown goodness.

Embrace the allure of raised-bed gardening, and let us cultivate a world where the seeds of dreams take root and blossom into a bountiful harvest of joy, nourishment, and personal fulfillment.

Chapter 1: Setting the Stage for Success

Choosing the Perfect Location: Sunlight and Soil Considerations

As you embark on your raised-bed gardening journey, the first crucial step is selecting an ideal location for your thriving garden. Sunlight plays a pivotal role in plant growth and development, so it's essential to choose a spot that receives ample sunlight throughout the day. Aim for at least six hours of direct sunlight per day, as most vegetables and flowers require this amount to flourish.

While sunlight is vital, soil quality is equally important for nurturing healthy plants. Ideally, your garden site should have well-draining soil that is rich in organic matter. Clay soil, with its dense texture, tends to retain water too well, while sandy soil drains too quickly, making it difficult to retain moisture and nutrients. Loamy soil, with its balanced composition of sand, silt, and clay, is considered ideal for raised-bed gardening.

Designing Your Raised-Bed Layout: Size and Shape Matters

The size and shape of your raised beds will depend on the available space in your garden and the types of plants you intend to grow. Larger raised beds are suitable for growing a variety of vegetables, while smaller beds can accommodate herbs, flowers, or smaller vegetables.

For rectangular raised beds, a common width is around 3-4 feet, as this allows for easy access from both sides without stepping on the soil. The length can vary depending on the available space, but it's generally recommended to keep it manageable to avoid overstretching when tending to the plants.

Square raised beds are another popular choice, offering a compact and efficient use of space. They are well-suited for smaller gardens or for growing a specific type of plant.

Selecting the Right Materials: Durability and Aesthetics

The materials you choose for constructing your raised beds will influence their durability, aesthetics, and overall functionality. Wood is a popular choice due to its natural beauty and ease of workability. However, it's important to select wood that is naturally rot-resistant, such as cedar or redwood, to ensure longevity.

Concrete blocks or bricks offer a more permanent solution, providing sturdy walls and a clean, modern aesthetic. However, they can be more expensive and require more effort to construct.

For a rustic touch, consider using galvanized metal sheets or corrugated roofing panels. These materials are durable and weather-resistant, but they may require additional support to prevent them from buckling.

Constructing Your Raised Beds: A Step-by-Step Guide

Once you've selected the location, designed your layout, and chosen the materials, it's time to embark on the exciting task of constructing your raised beds. Here's a step-by-step guide to help you get started:

1. Clear the Site: Remove any existing vegetation, debris, and rocks from the

chosen spot. Level the ground to ensure a stable base for your raised beds.

2. Mark the Layout: Use stakes and string to mark the outline of your raised beds, ensuring the dimensions are accurate and aligned with your design.

3. Cut the Materials: Cut the wood, concrete blocks, or metal sheets to the desired length and height of your raised beds.

4. Assemble the Walls: Secure the cut materials together using appropriate fasteners, such as nails, screws, or mortar. Ensure the walls are sturdy and level.

5. Line the Beds (Optional): To prevent soil erosion and water loss, consider lining your raised beds with landscaping fabric or hardware cloth.

6. Fill the Beds with Soil: Fill your raised beds with a mixture of topsoil, compost, and other organic matter to create a rich and fertile growing medium.

7. Water the Soil: Thoroughly water the newly filled raised beds to allow the soil to settle and establish a good moisture level.

With your raised beds in place, you're now ready to embark on the exciting journey of selecting and nurturing your plants, transforming your garden into a flourishing haven of homegrown goodness.

Chapter 2: Nurturing the Soil: The Foundation of Abundance

Understanding Soil Composition: The Essentials of Fertile Ground

Before delving into the world of plant selection and cultivation, it's essential to gain a fundamental understanding of soil composition, the bedrock of a thriving raised-bed garden. Soil is a complex ecosystem teeming with life, a living tapestry of minerals, organic matter, and countless microorganisms that work in harmony to support plant growth.

The three primary components of soil are sand, silt, and clay. Sand, with its coarse particles, provides drainage and aeration, while silt, with its fine particles, holds moisture and nutrients. Clay, with its even finer particles, retains water and nutrients but can become compacted if not managed properly.

The ideal soil for raised-bed gardening is a well-draining, loamy soil, a balanced blend of these three components. Loamy soil provides the right balance of drainage, moisture retention, and nutrient availability, creating an optimal environment for plant roots to thrive.

Creating a Nutrient-Rich Mix: Assembling the Elements of Success

While the natural soil in your garden may not be perfectly balanced, you can create a nutrient-rich mix that provides the ideal conditions for your plants to flourish. The key is to enrich the soil with organic matter, which acts as a sponge, holding moisture and nutrients, and provides a haven for beneficial microorganisms.

Compost, the decomposed remains of plant and animal matter, is a gardener's gold. It is a rich source of organic matter, humus, and essential nutrients, and it helps to improve soil structure, aeration, and drainage.

Other organic amendments, such as aged manure, leaf mold, and worm castings, can also be incorporated to enhance soil fertility. These amendments provide additional nutrients and beneficial microbes, further enriching the soil environment.

Enriching with Organic Matter: Compost and Other Amendments

Composting is a natural process that transforms organic waste into nutrient-rich humus, a valuable soil amendment. By composting kitchen scraps, yard trimmings, and other organic materials, you can create a sustainable source of nourishment for your raised beds.

To start a compost pile, simply choose a location that receives sunlight and has good drainage. Layer organic materials, alternating between green (nitrogen-rich) materials like grass clippings and kitchen scraps and brown (carbon-rich) materials like dry leaves and twigs.

Turn the compost pile regularly to aerate it and encourage decomposition. Within a few months, you should have a dark, crumbly compost ready to enrich your raised beds.

Maintaining Soil Health: Ongoing Care and Practices

Maintaining soil health is an ongoing process that requires regular attention and care. Throughout the growing season, there are several practices you can implement to ensure your raised beds remain fertile and supportive of plant growth.

Watering is crucial for maintaining soil moisture, but overwatering can lead to waterlogged conditions that harm plant roots. Water deeply and less frequently, allowing the soil to dry slightly between waterings.

Mulching, the practice of covering the soil surface with organic material, provides numerous benefits. It helps to retain soil moisture, suppress weeds, regulate soil temperature, and prevent erosion. Apply a layer of mulch, such as shredded leaves, straw, or bark, around your plants to keep the soil healthy and productive.

Soil testing is a valuable tool for assessing nutrient levels and identifying any potential imbalances. By testing your soil periodically, you can make informed decisions about adding appropriate amendments to maintain optimal soil fertility.

By incorporating these practices into your raised-bed gardening routine, you'll be well on your way to creating a thriving ecosystem that supports healthy, bountiful plants. Remember, the soil is the foundation of your garden, and nurturing it is an investment in the future success of your plants.

Chapter 3: Selecting Plants for Your Raised-Bed Garden

As you embark on the exciting journey of cultivating your raised-bed garden, the next crucial step is selecting the plants that will transform your plot of land into a flourishing haven of homegrown goodness. With a vast array of options available, choosing the right plants can be both exhilarating and daunting.

To make this process more manageable, let's explore the key considerations that will guide your plant selection.

Exploring Plant Varieties: Choosing the Right Options for Your Climate

Your geographical location and climate will play a significant role in determining the types of plants you can successfully grow in your raised beds. Familiarize yourself with your hardiness zone, a classification system that divides the country into regions based on average annual minimum temperatures.

Choose plants that are well-suited to your hardiness zone to ensure their survival and optimal growth throughout the year. Consult local nurseries, gardening centers, or online resources to identify plants that thrive in your specific climate.

Understanding Plant Spacing: Optimal Growth and Yield

Each plant species has its unique growth habit and spacing requirements. Proper spacing allows plants to receive adequate sunlight, air circulation, and room to spread their roots, all of which are essential for healthy growth and bountiful harvests.

Overcrowded plants compete for resources, leading to stunted growth, reduced yields, and increased susceptibility to pests and diseases. Conversely, excessive spacing can result in wasted space and reduced productivity.

Refer to plant tags or online resources to determine the appropriate spacing for each type of plant you intend to grow. Remember, a well-spaced garden is a happy and productive garden.

Planning for Succession Planting: A Continuous Harvest

Extend the bounty of your raised-bed garden by incorporating succession planting, a technique that involves planting a series of crops throughout the growing season to ensure a continuous harvest.

Plan your succession planting schedule by considering the maturity dates of different vegetable varieties. Stagger planting times to ensure a steady supply of fresh produce from spring through fall.

Companion Planting: Enhancing Growth and Repelling Pests

Discover the wonders of companion planting, a time-honored practice that combines certain plant species to enhance growth, repel pests, and attract beneficial insects.

For instance, interplanting rows of onions and carrots can deter carrot flies, while planting nasturtiums near squash can repel squash bugs. Experiment with companion planting combinations to create a harmonious and pest-resistant garden ecosystem.

With careful consideration of climate, spacing, succession planting, and companion planting, you'll be well-equipped to select the ideal plants for your raised-bed garden, transforming your outdoor space into a thriving haven of homegrown goodness.

Chapter 4: Cultivation and Care: Nurturing Your Plants to Thriving

Now that your raised beds are brimming with carefully selected plants, it's time to embark on the nurturing journey of cultivation and care. Just as a newborn child requires continuous attention and nourishment, your plants will thrive with your dedicated care.

Watering Techniques: Providing Life-Giving Moisture

Watering is the lifeblood of your garden, providing the essential moisture that plants need for photosynthesis, nutrient uptake, and cell growth. However, overwatering and underwatering can both harm your plants.

The frequency and depth of watering depend on several factors, including the type of plants, weather conditions, and soil composition. As a general rule of thumb, water deeply and less frequently, allowing the soil to dry slightly between waterings.

Weed Management: Maintaining a Clean and Healthy Environment

Weeds are unwanted plant competitors that can steal precious resources from your desired crops. Regular weed control is essential for maintaining a healthy and productive raised-bed garden.

Hand-pulling weeds is an effective method for small infestations. For larger areas, consider using a hoe or cultivator to loosen the soil and uproot weeds. Organic mulches, such as shredded leaves or straw, can also help suppress weed growth.

Fertilization and Mulching: Sustaining Growth and Enhancing Yield

Fertilization provides the essential nutrients that plants need to grow and produce abundant harvests. Organic fertilizers, such as compost, aged manure, and fish emulsion, are excellent choices for raised-bed gardening due to their slow-release nature and ability to improve soil health.

Mulching, as mentioned previously, not only helps with weed control but also offers additional benefits, such as retaining soil moisture, regulating soil temperature, and preventing erosion. Apply a layer of mulch around your plants to keep the soil healthy and productive.

Pest and Disease Control: Protecting Your Precious Plants

Despite your best efforts, your raised-bed garden may encounter pests and diseases that can damage your plants and reduce yields. Regular pest and disease monitoring is crucial for early detection and timely intervention.

To minimize the risk of pest infestations, encourage beneficial insects, such as ladybugs and lacewings, by planting attractive flowers and avoiding the use of broad-spectrum pesticides.

For disease prevention, practice good sanitation by removing diseased plants and debris, and rotate your vegetable crops to avoid building up soil-borne diseases.

With consistent watering, weed control, fertilization, mulching, and pest and disease management practices, you'll be well on your way to cultivating a thriving raised-bed garden that rewards you with bountiful harvests and a sense of accomplishment. Remember, your plants are like living extensions of your care, and their flourishing growth is a testament to your dedication.

Chapter 5: Harvesting and Preserving: Enjoying the Fruits of Your Labor

As the vibrant hues of ripening fruits and vegetables adorn your raised beds, the anticipation of harvest fills the air. The moment has arrived to reap the rewards of your dedication, to gather the fruits of your labor and savor the bounty of your garden.

Recognizing Harvest Readiness: Timing Matters

Harvesting at the peak of ripeness ensures the best flavor, texture, and nutritional value of your produce. Different plant species have varying maturity dates, so it's essential to be familiar with the timing of each crop.

Observe the visual cues of ripeness, such as color changes, softening texture, and distinctive aromas. Some vegetables, like cucumbers and beans, are best harvested while still young and tender, while others, like tomatoes and melons, require more time to develop their full flavor and sweetness.

Harvesting Techniques: Gathering Your Bountiful Harvest

Each type of plant requires a specific harvesting technique to minimize damage and ensure future productivity. Gently grasp leaves and herbs, using scissors or a sharp knife for clean cuts.

For fruits, select ripe specimens, using gentle pressure to detach them from the plant. With root vegetables, carefully loosen the soil around the base and gently pull them up, taking care not to damage the roots.

Preservation Methods: Savoring Your Garden's Goodness

Preserve the abundance of your harvest to enjoy its goodness throughout the year. Freezing is a versatile method suitable for a variety of fruits, vegetables, and herbs. Blanching, a quick pre-cooking step, helps preserve color, texture, and nutrients.

Drying is an excellent method for preserving herbs, fruits, and some vegetables. Sun-drying is a traditional method, while dehydrators offer a more controlled drying process.

Canning, while requiring more time and effort, allows for long-term storage of fruits, vegetables, and even jams and sauces. Follow safe canning procedures to ensure the preservation of freshness and prevent foodborne illnesses.

Sharing the Abundance: Gifts from Your Garden

The joy of gardening extends beyond personal enjoyment; it's an opportunity to share the abundance with loved ones and community members. Share your harvest by gifting fresh produce, creating homemade jams or jellies, or hosting a garden-to-table gathering.

Remember, gardening is not just about cultivating plants; it's about cultivating connections, fostering a sense of community, and sharing the simple pleasures that nature provides.

As you savor the fruits of your labor, let the satisfaction of your harvest be a reminder of the power of nurturing, the beauty of growth, and the joy of sharing nature's bounty.

Chapter 6: Enhancing Your Raised-Bed Gardening Experience

As you delve deeper into the world of raised-bed gardening, you'll discover a wealth of opportunities to enhance your experience, transforming your garden into a haven of not only bountiful harvests but also creativity, personal fulfillment, and a deeper connection with nature.

Extending the Growing Season: Utilizing Season Extenders

Mother Nature may have her set seasons, but you don't have to be bound by them. With the help of season extenders, you can extend your growing season and enjoy fresh produce even when the weather outside is frightful.

Cloches, clear glass or plastic enclosures placed over plants, protect tender seedlings from spring frosts and extend their growing season. Row covers, made of fabric or plastic, provide similar protection and can also help retain heat and moisture.

Cold frames are raised structures with transparent covers, providing a sheltered environment for growing vegetables even in cooler months. Hotbeds, heated from below, offer even more warmth, allowing you to start seeds earlier and enjoy an extended harvest.

Incorporating Vertical Gardening: Maximizing Space Utilization

Especially if your gardening space is limited, vertical gardening offers an ingenious solution to maximize space utilization and create a visually striking garden.

Trellises and arbors provide support for climbing plants, such as beans, tomatoes, and cucumbers, allowing them to grow upwards instead of outwards. This not only saves valuable ground space but also enhances the aesthetics of your garden.

Hanging planters, containers suspended from above, are another excellent way to utilize vertical space. Herbs, flowers, and even small vegetables can thrive in hanging planters, adding a touch of greenery and vibrancy to your garden.

Embracing Creative Design: Personalizing Your Garden Aesthetic

Your raised-bed garden is not just a place for growing plants; it's an expression of your personal style and creativity. Take the opportunity to design a garden that reflects your unique tastes and preferences.

Incorporate decorative elements, such as garden statues, birdbaths, or colorful pots, to add a touch of whimsy and charm. Experiment with different colors and textures of plants to create a visually appealing tapestry.

Consider incorporating pathways made of gravel, stepping stones, or wooden planks, providing easy access to your plants and adding a touch of elegance to your garden layout.

Attracting Beneficial Insects: Fostering a Thriving Ecosystem

Your raised-bed garden can become a haven for beneficial insects, the unsung heroes of a healthy ecosystem. These insects play a crucial role in pollination, pest control, and nutrient cycling, ensuring a thriving garden.

Plant flowers that attract pollinators, such as bees and butterflies. Avoid using pesticides that could harm these beneficial insects. Provide nesting sites for native bees and other beneficial insects by creating small piles of rocks or logs.

By fostering a welcoming environment for beneficial insects, you'll create a more balanced and resilient garden ecosystem, reducing your reliance on artificial pest control methods and promoting natural harmony in your raised-bed haven.

As you embark on these enriching endeavors, remember that raised-bed gardening is a continuous journey of learning, experimentation, and personal growth. Embrace the challenges, celebrate the successes, and cherish the bountiful rewards that this rewarding pursuit has to offer.

Conclusion: Celebrating the Joys of Raised-Bed Gardening

As you stand amidst your thriving raised-bed garden, a sense of accomplishment washes over you. The once barren patch of land has transformed into a flourishing oasis, teeming with life and brimming with the promise of bountiful harvests.

With each passing season, you've witnessed the miraculous cycle of growth, from the tiny seeds that held the potential for life to the vibrant plants that now grace your garden. You've learned the delicate balance of nature, the importance of nurturing and care, and the profound satisfaction of cultivating something from seed to harvest.

Raised-bed gardening has not only enriched your table with fresh, wholesome produce but also enriched your life with a deeper connection to nature, a newfound appreciation for the intricacies of plant life, and a sense of accomplishment that stems from nurturing something from seed to harvest.

As you reflect on your gardening journey, you realize that it's not just about growing plants; it's about cultivating patience, perseverance, and a deep appreciation for the natural world. It's about fostering harmony with nature, learning from its wisdom, and celebrating the cycles of life that unfold within your raised-bed haven.

Embrace the continuous learning process that lies ahead. Experiment with new plant varieties, explore different gardening techniques, and share your knowledge and enthusiasm with others. Remember, gardening is a journey, not a destination.

As you continue to nurture your raised-bed garden, may you find joy in the simple pleasures of watching plants grow, savor the flavors of homegrown goodness, and share the abundance with those around you. For within the heart of your garden lies a world of beauty, wonder, and endless possibilities.